CCEA | GCSE
FURTHER MATHS REVISION BOOKLET
STATISTICS

COLOURPOINT EDUCATIONAL

Contents

Revision Exercise 1 3

Revision Exercise 2 10

Revision Exercise 3 17

Revision Exercise 4 23

Revision Exercise 5 29

Answers 36

Neill Hamilton

CCEA GCSE
FURTHER MATHS REVISION BOOKLET
STATISTICS

COLOURPOINT EDUCATIONAL

Name:

© Neill Hamilton and Colourpoint Creative Ltd 2021

ISBN: 978 1 78073 319 7

First Edition
Second Impression 2025

Layout and design: April Sky Design
Printed by: GPS Colour Graphcs Ltd, Belfast

All rights reserved. No part of this publication may be reproduced, stored in a retrieval system or transmitted in any form or by any means, electronic, mechanical, photocopying, scanning, recording or otherwise, without the prior written permission of the copyright owners and publisher of this book.

The Author

Neill Hamilton will be well known to Mathematics teachers in Northern Ireland. Until his retirement in 2012, he was a teacher of GCSE Mathematics and Additional/Further Mathematics at a Northern Ireland comprehensive school. His previous publications include *Further Mathematics for CCEA GCSE*, and GCSE Mathematics Revision Booklets *M3* and *M4*, also published by Colourpoint.

Dedicated to Arlene, for everything she has done for me, and to Marley who is the best and most loyal friend I could ever have.

Colourpoint Educational
An imprint of Colourpoint Creative Ltd
Colourpoint House
Jubilee Business Park
21 Jubilee Road
Newtownards
County Down
Northern Ireland
BT23 4YH

Tel: 028 9182 0505
E-mail: sales@colourpoint.co.uk
Web site: www.colourpointeducational.com

Publisher's Note: This book has been written to help students preparing for the GCSE Further Mathematics specification from CCEA. While Colourpoint Educational and the authors have taken every care in its production, we are not able to guarantee that the book is completely error-free. Additionally, while the book has been written to closely match the CCEA specification, it is the responsibility of each candidate to satisfy themselves that they have fully met the requirements of the CCEA specification prior to sitting an exam set by that body. For this reason, and because specifications change with time, we strongly advise every candidate to avail of a qualified teacher and to check the contents of the most recent specification for themselves prior to the exam. Colourpoint Creative Ltd therefore cannot be held responsible for any errors or omissions in this book or any consequences thereof.

Revision Exercise 1

1. Catrina recorded the number of swimming lengths in metres that she swam each day in April.

Length in metres	Number of lengths		
14	7		
15	9		
16	5		
17	8		
18	1		

   Calculate the:
   (a) mean,

   Answer _____ [2]

   (b) standard deviation,

   Answer _____ [2]

2. There are 12 boys and 8 girls in a class.
   The boys have a mean height of 1.48 m and a standard deviation of 1.6 m.
   The girls have a mean height of 1.37 m and a standard deviation of 1.74 m.
   Calculate the
   (a) mean of the heights of all the pupils in the class,

   Answer _____ [3]

(b) standard deviation of the heights of all the pupils in the class,

Answer _____ [3]

3. The number of cards, labelled with letters, is shown in the table below.

Letter on a card	A	E	I	O	U
Number of cards	5	4	8	2	1

Steven takes 2 cards at random, without replacement.
(a) Work out the probability that both cards are labelled A or both cards are labelled E.

Answer _____ [1]

He then takes a third card at random.
(b) Work out the probability that all three cards are labelled A or all three cards are labelled E.

Answer _____ [1]

(c) Given that both the first two cards are labelled A or both labelled E, work out the probability that all three cards are labelled A or all three cards are labelled E.

Answer _____ [2]

4. The table below shows the heights and weights of 9 objects.

Height (cm)	26	22	27	19	22	23	29	18	30
Weight (kg)	4.8	4.1	4.4	3.7	3.8	4.3	5.2	3.5	4.9
Rank (Height)									
Rank (Weight)									

(a) Write down the rank orders for the heights and weights in the table.

[2]

(b) Calculate Spearman's coefficient of rank correlation.

Answer _____ [2]

(c) Interpret your answer to part (b).

Answer _____ [1]

(d) Calculate the mean height and weight.

Answer _____ [1]

The data from the table are plotted on the graph below.

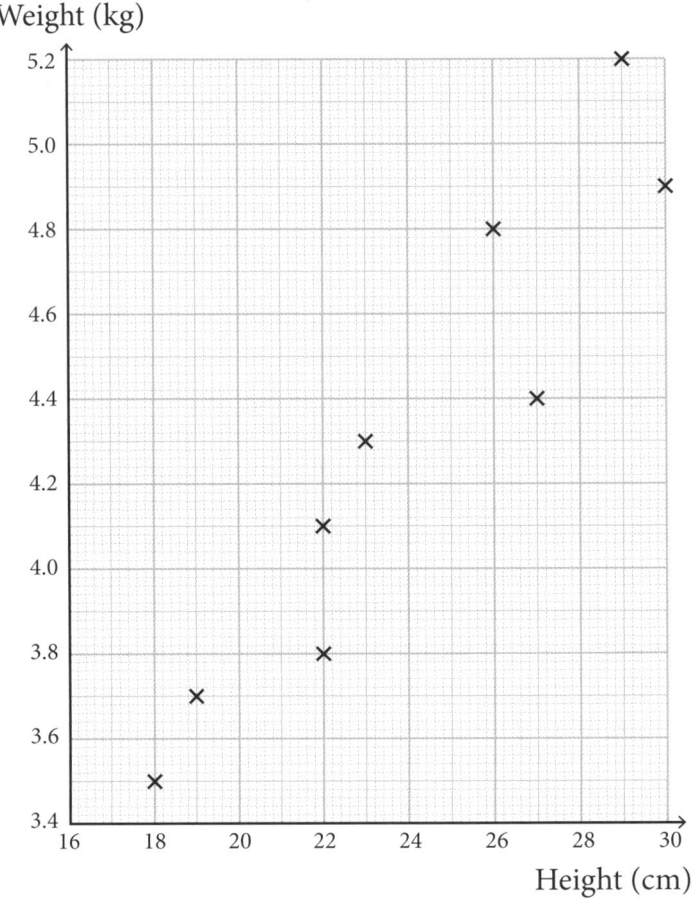

(e) Draw your line of best fit on the graph.

[2]

(f) Determine the equation of this line of best fit.

Answer _____ [3]

5. There were 116 guests for breakfast at a hotel:
   42 chose sausages,
   11 chose sausages and eggs,
   4 chose bacon, sausages and eggs,
   19 chose only sausages,
   15 chose only bacon,
   18 chose no bacon, no sausages and no eggs.

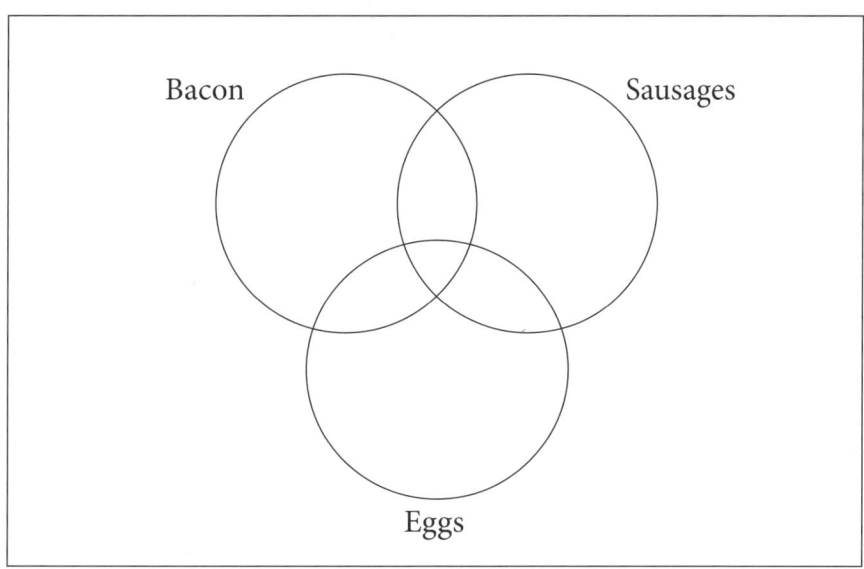

(a) Using the Venn diagram, find the probability that:
 (i) a guest, selected at random, chose bacon and sausages;

Answer _____ [3]

(ii) a guest, selected at random, chose eggs.

Answer _____ [1]

(b) Given that a guest, selected at random, chose eggs find the probability that the guest chose sausages.

Answer _____ [2]

6. The heights of a set of objects are normally distributed with mean 37.4 cm and standard deviation 5.8 cm. Find the probability that an object, chosen at random, is:
   (a) higher than 45.52 cm,

   Answer _____ [4]

   (b) higher than 24.93 cm.

   Answer _____ [2]

7. (a) Using Pascal's triangle, or otherwise, write out the expansion of $(p + q)^6$

   Answer _____ [2]

(b) The probability that it rains each day is 0.64.
Over a period of six days find the probability that:
(i) it rains on two days,

Answer _____ [3]

(ii) it rains on at least three days.

Answer _____ [3]

8. There are *x* bungalows and 10 detached houses in a development.
There are more bungalows than detached houses.
Two properties are chosen at random.
The probability that a bungalow and a detached house are chosen is $10/21$.
Work out the probability that 2 bungalows are chosen.

Answer _____ [5]

Total for revision exercise [50]

Revision Exercise 2

1. Connor recorded the number of points scored by 30 people in a game.

Number of points	Number of people			
1–5	3			
6–14	14			
15–20	9			
21–22	4			

   Calculate the:
   (a) mean,

   Answer _____ [2]

   (b) standard deviation.

   Answer _____ [3]

2. A bookshop sold 28 hardback books and 12 paperback books yesterday.
   The hardback books had a mean cost of £11 and a standard deviation of £2.50
   All the books had a mean cost of £9.60 and a standard deviation of £3.20
   Calculate the:
   (a) mean,

   Answer _____ [2]

(b) standard deviation of the costs of the paperback books.

Answer _____ [3]

3. There are 16 dogs and 12 cats in Marley's Boarding Kennels.
   The probability that an animal chosen at random in Marley's Boarding Kennels is under 6 years old is ¾.
   There are 13 dogs under 6 years old.
   (a) Calculate the probability that a cat chosen at random is under 6 years old.

Answer _____ [3]

An animal is chosen at random in Marley's Boarding Kennels.
(b) Calculate the probability that the animal is a dog, given that the animal chosen is at least 6 years old.

Answer _____ [1]

4. The table below shows the ages and costs of 8 cars in a garage.

Age (years)	7	5	9	14	10	15	9	9
Cost (£)	13 000	14 200	8 600	7 000	9 000	6 400	11 000	10 400
Rank (Age)								
Rank (Cost)								

(a) Write down the rank orders for the ages and costs in the table.

[2]

(b) Calculate Spearman's coefficient of rank correlation.

Answer _____ [2]

(c) Interpret your answer to part (b).

Answer _____ [1]

(d) Calculate the mean age and cost.

Answer _____ [1]

The data from the table are plotted on the graph below.

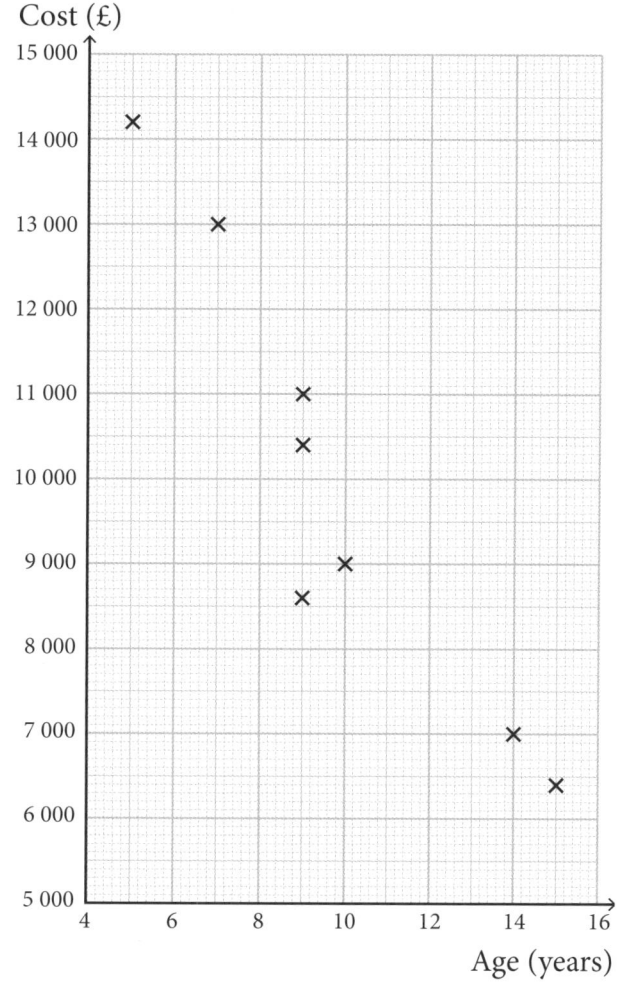

(e) Draw your line of best fit on the graph.

[2]

(f) Determine the equation of this line of best fit.

Answer _____ [3]

5. A sample of 76 people were asked whether they listened to Jazz, Rock and/or Classical music
33 said they listened to Jazz,
44 said they listened to Rock,
4 said they listened to Jazz, Rock and Classical,
12 said they listened to Jazz and Classical,
5 said they listened to none of these types of music.
The probability that a person, selected at random, listened to Rock given that the person listened to Classical was $7/22$.

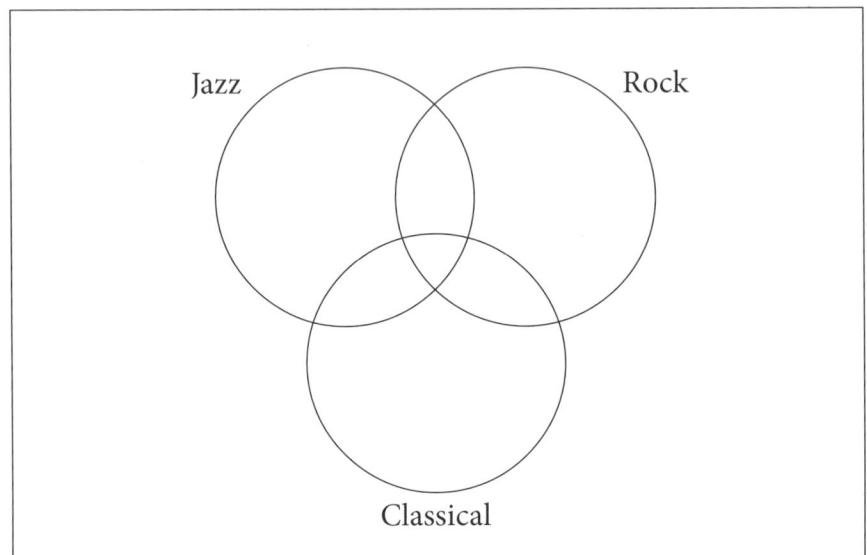

(a) Using the Venn diagram find the probability that a person, selected at random, listened to only Jazz.

Answer _____ [3]

(b) Given that a person, selected at random, listened to Rock find the probability that the person listened to Classical.

Answer _____ [2]

6. The marks in an exam are normally distributed with mean 58 and standard deviation 8.
   The marks are graded C, B and A.
   The highest mark getting a grade C is 52.
   The highest mark getting a grade B is 69.
   Find the probability that an pupil, chosen at random, gets a:
   (a) grade C,

   Answer _____ [4]

   (b) grade A.

   Answer _____ [2]

7. (a) Using Pascal's triangle, or otherwise, write out the expansion of $(p + q)^7$

   Answer _____ [2]

   (b) The probability of getting heads when tossing a biassed coin is $3/5$.
   7 coins are tossed.
   Find the probability of getting:
   (i) no heads,

   Answer _____ [3]

(ii) more heads than tails.

Answer _____ [3]

8. There are *x* laptops and 12 tablets for sale.
The probability that at least one laptop is sold in the first two sales is $62/95$.
Work out the probability that two laptops are sold in the first two sales.

Answer _____ [6]

Total for revision exercise [50]

Revision Exercise 3

1. Cadence recorded the number of texts sent in one day by people in a youth club.

Number of texts	Number of people		
12	7		
14	8		
17	12		
20	x		
22	11		
26	4		
29	2		
36	1		

The mean number of texts sent was 19.

(a) Find the value of x.
A solution by trial and improvement is not allowed.

Answer _____ [2]

(b) Work out the standard deviation.

Answer _____ [2]

2. The mean and standard deviation of the times taken by runners to complete a 5K race were 18.7 minutes and 2.6 minutes.
The runners are to run a 10K next week.
Each of their 5K times were doubled and a further 3 minutes added to each 5K time to give a predicted 10K time.
Find the:
(a) mean,

Answer _____ [2]

(b) standard deviation of the predicted 10K times.

Answer _____ [2]

3. The probability that a train arriving in Newry is full is 0.24.
   The probability that a train arriving in Newry is on time given that it is full is 0.56.
   The probability that a train arriving in Newry is on time is 0.6.
   Work out the probability that:
   (a) a train arriving in Newry is on time and full,

Answer _____ [2]

   (b) a train arriving in Newry is full given that it is on time.

Answer _____ [2]

4. The table below shows the volumes and masses of 8 objects.

Volume (cm³)	6	8	5	9	11	12	8	13
Mass (grams)	62	86	64	92	120	138	94	128
Rank (Volume)								
Rank (Mass)								

(a) Write down the rank orders for the volumes and masses in the table. [2]
(b) Calculate Spearman's coefficient of rank correlation.

Answer _____ [2]

(c) Interpret your answer to part (b).

Answer _____ [1]

(d) Calculate the mean volume and mass.

Answer _____ [1]

The data from the table are plotted on the graph below.

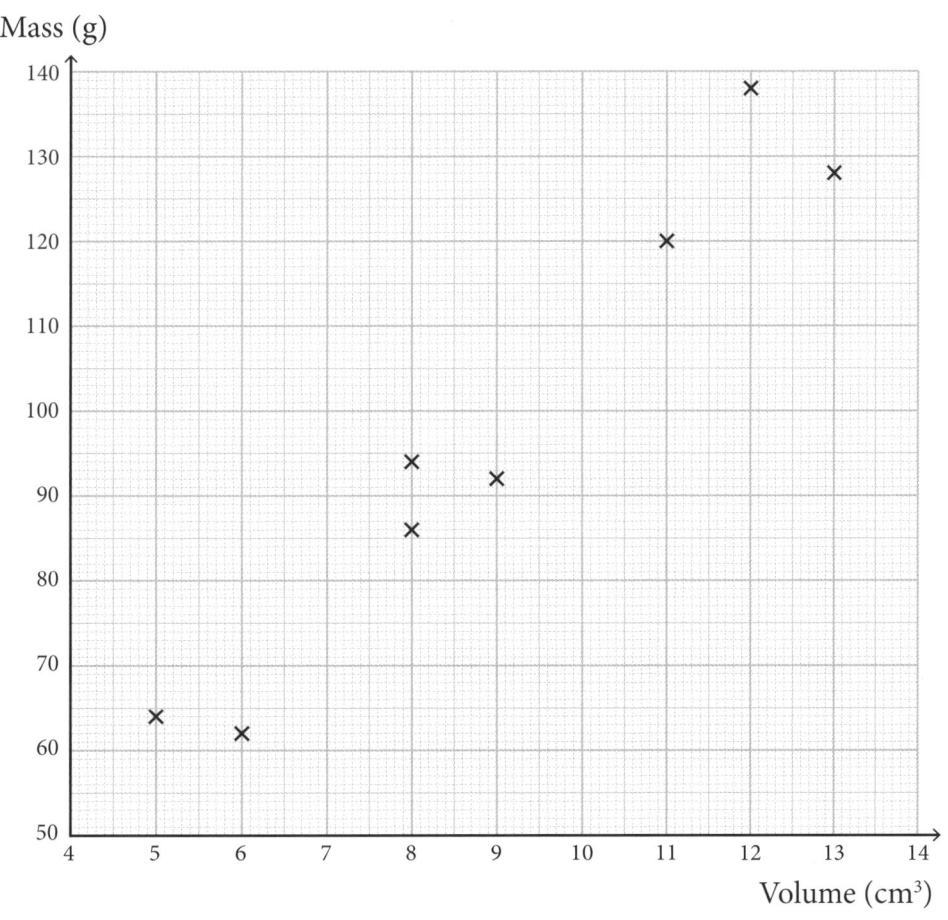

(e) Draw your line of best fit on the graph.

[2]

(f) Determine the equation of this line of best fit.

Answer _____ [3]

5. In training for a marathon David runs either a Long run or a Short run each day.
   On Monday the probability that David ran a Long run was 0.74
   The probability that David ran a Long run on both Monday and Tuesday was 0.2072
   (a) Using a tree diagram, or otherwise, work out the probability that David ran a Long run on Monday and a Short run on Tuesday.

   Answer _____ [3]

   The probability that David ran a Long run and a Short run on these two days was 0.7512
   (b) Work out the probability that David ran 2 Short runs on Monday and Tuesday.

   Answer _____ [3]

6. The ages of a group of doctors are normally distributed with mean 42.912 years and standard deviation 2.8 years.
   Find the probability that a doctor, chosen at random, is:
   (a) older than 44

   Answer _____ [4]

(b) at most 47

Answer _____ [2]

7. **(a)** Using Pascal's triangle, or otherwise, write out the expansion of $(p + q)^4$

Answer _____ [2]

(b) There are 4 red, 2 yellow and 10 green counters in a bag.
4 counters are chosen at random.
Find the probability of choosing:
(i) at least 1 green,

Answer _____ [3]

(ii) 1 yellow or 2 red counters.

Answer _____ [3]

8. A group of 74 tourists were asked which places they had visited in Belfast.
   3 had visited the Zoo, the Titanic and the Museum,
   8 had visited the Zoo and the Museum,
   11 had visited the Titanic and the Museum,
   32 had visited the Museum,
   27 had visited the Titanic,
   ⅔ of those who had visited the Zoo had not visited either the Titanic or the Museum,
   2 had not visited the Zoo, the Titanic or the Museum.

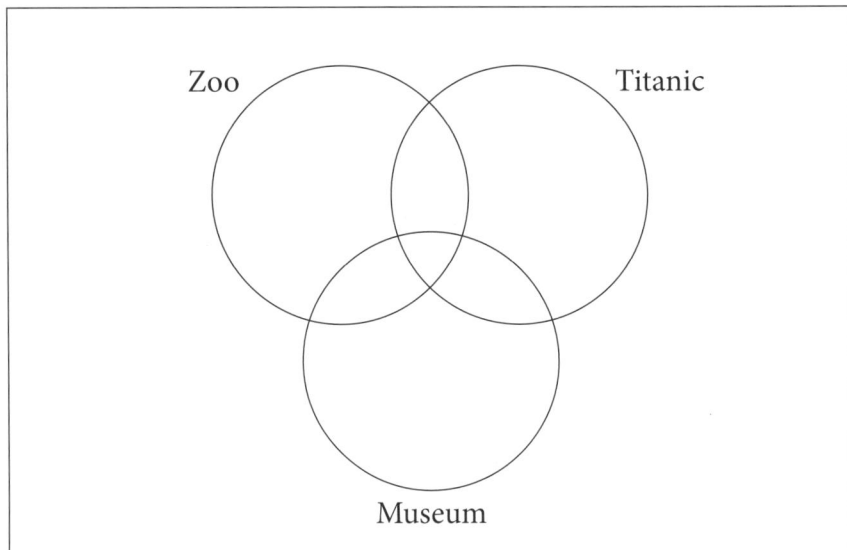

Use the Venn Diagram to find the probability that a tourist visited the Zoo given that she visited the Titanic.

Answer _____ [7]

Total for revision exercise [50]

Revision Exercise 4

1. Colin recorded the weights, W kg, of plants for sale.

Weight W kg	Number		
$1 < W \leq 6$	6		
$6 < W \leq 14$	15		
$14 < W \leq 17$	22		
$17 < W \leq 25$	4		
$25 < W \leq A$	3		

   The mean weight was 13.6 kg.

   (a) Find the value of A.
   A solution by trial and improvement is not allowed.

   Answer _____ [2]

   (b) Work out the standard deviation.

   Answer _____ [2]

2. The mean and standard deviation of the test scores of pupils was 54 and 4
   The scores were then scaled so that the mean became 47 and the standard deviation was halved by multiplying each score by a constant p and adding a constant q.
   Work out the value of
   (a) p

   Answer _____ [1]

(b) q

Answer _____ [2]

3. Gerwyn plays darts.
   The probability he scores 60 with his first throw is 0.47
   The probability he scores 60 with his second throw is 0.58
   The probability he scores 60 with both his first two throws is 0.24
   Work out the probability that he does not score 60 with his first or second throw.

Answer _____ [4]

4. The table below shows the engine size in litres of cars and the distance travelled in km on one litre of petrol for each car.

Size (litres)	0.8	2.4	1.2	1.6	1.2	2.8	3.2	2.8
Distance (km)	13.2	6.4	14.8	8.8	14.4	8.4	4	7.6
Rank (Size)								
Rank (Distance)								

(a) Write down the rank orders for the sizes and distances in the table. [2]

(b) Calculate Spearman's coefficient of rank correlation.

Answer _____ [2]

(c) Interpret your answer to part (b).

Answer _____ [1]

(d) Calculate the mean size and distance.

Answer _____ [1]

The data from the table are plotted on the graph below.

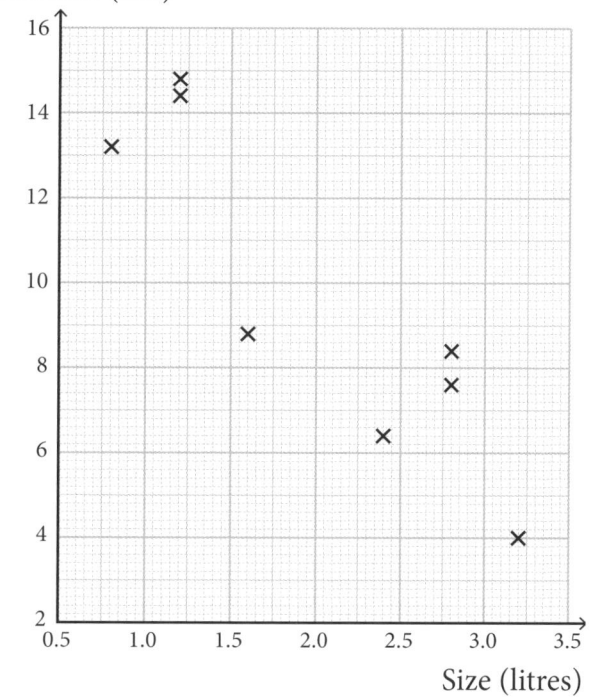

(e) Draw your line of best fit on the graph.

[2]

(f) Determine the equation of this line of best fit.

Answer _____ [3]

5. The probability that a bus is on time any day is 5/9.
   If it is on time the probability that I get the bus is 2/5.
   If it is not on time the probability that I get the bus is 9/16.

   (a) Draw a clearly labelled tree diagram to show **all** possible probabilities.

   Answer _____ [2]

   (b) Work out the probability that I get the bus.

   Answer _____ [2]

   (c) Work out the probability that I miss the bus given that it is on time.

   Answer _____ [1]

   (d) I get the bus. Work out the probability that it is not on time.

   Answer _____ [2]

Revision Exercise 4

(e) Work out the probability that I get 2 buses in 3 days.

Answer _____ [2]

6. The weights of a set of objects are normally distributed with mean 2.3 kg and standard deviation 520 g. Find the probability that an object, chosen at random, is:

(a) heavier than 2.2 kg,

Answer _____ [4]

(b) at least 2.76 kg.

Answer _____ [2]

7. (a) Using Pascal's triangle, or otherwise, write out the expansion of $(p + q)^8$

Answer _____ [2]

(b) Lily plays Quinn 8 games of chess.
The probability Lily wins each game is 0.52
Find the probability that Lily:
(i) wins 2 games,

Answer _____ [2]

(ii) wins at most 2 games.

Answer _____ [2]

8. The frequency distribution of the scores below has a standard deviation of $\sqrt{14}$
Work out the value of n.

Score	Frequency		
1	4		
2	3		
3	2		
n	1		

Answer _____ [7]

Total for revision exercise [50]

Revision Exercise 5

1. Niamh recorded the lengths, l cm, of some sticks.

Length l cm	Number		
$1 < l \leq 4$	2		
$4 < l \leq 7$	6		
$7 < l \leq 10$	5		
$10 < l \leq 13$	4		
$13 < l \leq 16$	x		

   The mean length was 8.5 cm.
   (a) Find the value of x.
   A solution by trial and improvement is not allowed.

   Answer _____ [2]

   (b) Work out the standard deviation.

   Answer _____ [2]

2. The mean time 10 people spent queuing at a takeaway was 12 minutes and the standard deviation was 2 minutes.
   David and Carly then queued, with Carly queuing twice as long as David.
   The mean time all 12 people spent queuing was 11.5 minutes.
   (a) Calculate how long:
   (i) David queued,

   Answer _____ [3]

(ii) Carly queued.

Answer _____ [1]

(b) Calculate the standard deviation for all 12 people.

Answer _____ [2]

3. There are 115 members of a youth club on a weekend activity break.
   62 went cycling,
   46 went swimming,
   18 did not cycle or swim,
   Use a Venn diagram, or otherwise, to:
   (a) work out the probability that a member, selected at random, went cycling and swimming,

Answer _____ [2]

(b) work out the probability that a member, selected at random, went swimming given that this member went cycling.

Answer _____ [1]

4. The table below shows the time taken by 10 workers to drive home from work and the mean speeds.

Time (minutes)	49	48	34	45	31	42	43	35	37	39
Speed (mph)	38	52	63	44	71	52	34	52	56	74
Rank (Time)										
Rank (Speed)										

(a) Write down the rank orders for the times and speeds in the table.

[2]

(b) Calculate Spearman's coefficient of rank correlation.

Answer _____ [2]

(c) Interpret your answer to part (b).

Answer _____ [1]

(d) Calculate the mean time and speed.

Answer _____ [1]

The data from the table are plotted on the graph below.

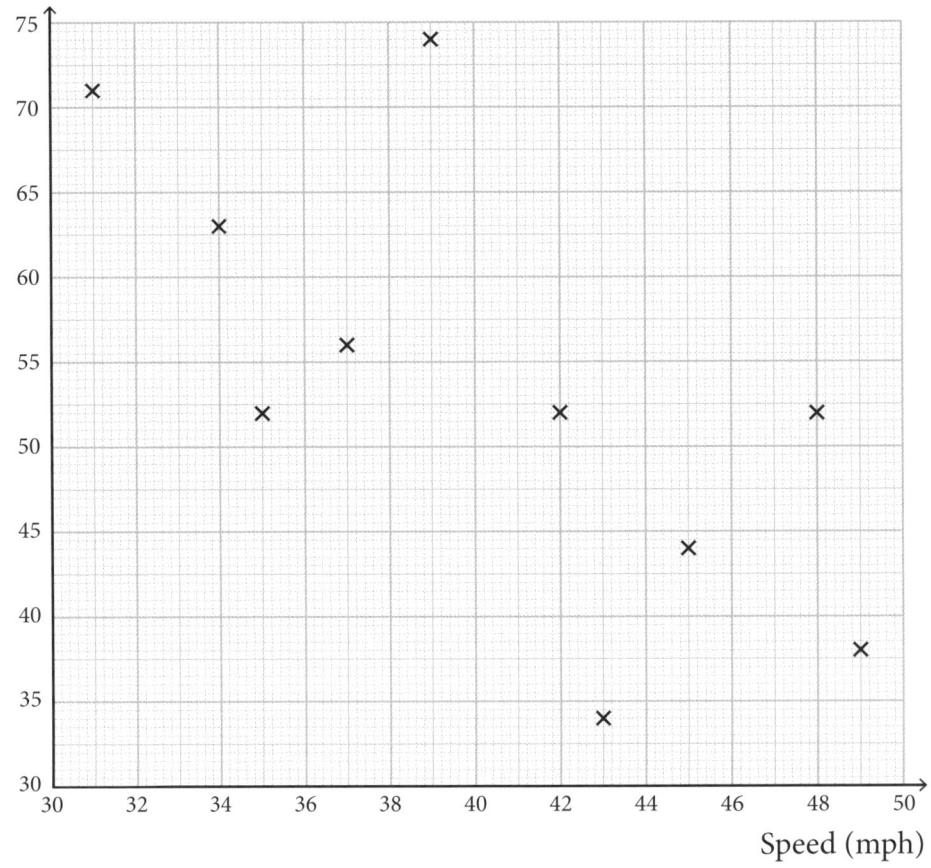

(e) Draw your line of best fit on the graph.

[2]

(f) Determine the equation of this line of best fit.

Answer _____ [3]

5. The pupils in a class were asked how they came to school.
8 came by bus,
4 walked,
16 came by car.
Two pupils were chosen at random.
   (a) Using a tree diagram, or otherwise, calculate the probability that:
      (i) both came by the same method,

Answer _____ [3]

      (ii) at least one walked.

Answer _____ [2]

   (b) Given that the first pupil walked to school calculate the probability that the second pupil came by car.

Answer _____ [1]

6. The times taken by runners in a marathon are normally distributed with mean 3 ½ hours and standard deviation 24 minutes.
   Find the probability that an runner, chosen at random, ran in:
   (a) less than 4 hours,

   Answer _____ [4]

   (b) less than 2 hours 25 minutes.

   Answer _____ [2]

7. (a) Using Pascal's triangle, or otherwise, write out the expansion of $(p + q)^5$

   Answer _____ [2]

   When Jim comes home from school the first thing he does is either watch TV, play outside or do his homework.
   The probability he watches TV each day is 0.61.
   He is twice as likely to play outside each day rather than to do homework.
   (b) Find the probability, in a full school week of 5 days, that the first thing Jim does:
       (i) is to watch TV each day,

   Answer _____ [2]

(ii) is to play outside on 2 of the days.

Answer _____ [3]

(c) Show that it is almost a 50/50 chance that Jim does his homework first on at least one of the days.

Answer _____ [2]

8. Rachel recorded the number of lengths swam by boys and girls.
Most of the data is shown in the two tables below.

Boys

Number of lengths	Number of boys
5	
10	17
15	14
20	9
25	8

Girls

Number of lengths	Number of girls
5	
10	18
15	15
20	12
25	18

Six times as many girls swam 5 lengths as boys.
The mean number of lengths swum was the same for both boys and girls.
Calculate the number of girls who swam 5 lengths.

Answer _____ [5]

Total for revision exercise **[50]**

Answers

Note: Due to rounding, follow-on answers may vary in the last significant figure throughout.

Revision Exercise 1

1.

Length in metres (x)	Number of lengths (f)	fx	fx^2
14	7	98	1372
15	9	135	2025
16	5	80	1280
17	8	136	2312
18	1	18	324
Total	30	467	7313

(a) $467 \div 30 = 15.57$ [M1W1]
(b) $\sqrt{(7313 \div 30) - 15.57^2} = 1.16$ [M1W1]

2. (a) Boys: $12 \times 1.48 = 17.76$; Girls $8 \times 1.37 = 10.96$
Total $= 17.76 + 10.96 = 28.72$; Mean $= 28.72 \div 20$
$= 1.436$ m [M1W2]
(b) Boys: $1.6 = \sqrt{(\Sigma x^2 \div 12) - 1.48^2}$, so
$\Sigma x^2 = 57.0048$; Girls: $1.74 = \sqrt{(\Sigma x^2 \div 8) - 1.37^2}$, so
$\Sigma x^2 = 39.236$; Total $\Sigma x^2 = 57.0048 + 39.236$
$= 96.2408$; $s = \sqrt{(96.2408 \div 20) - 1.436^2} = 1.66$
[M1W2]

3. (a) $5/20 \times 4/19 + 4/20 \times 3/19 = 8/95$ [1]
(b) $5/20 \times 4/19 \times 3/18 + 4/20 \times 3/19 \times 2/18 = 7/570$ [1]
(c) $7/570 \div 8/95 = 7/48$ [1]

4. (a) [MW2]

Height (cm)	26	22	27	19	22	23	29	18	30
Weight (kg)	4.8	4.1	4.4	3.7	3.8	4.3	5.2	3.5	4.9
Rank (Height)	6	3.5	7	2	3.5	5	8	1	9
Rank (Weight)	7	4	6	2	3	5	9	1	8
d	1	0.5	−1	0	−0.5	0	1	0	−1
d^2	1	0.25	1	0	0.25	0	1	0	1

(b) $\Sigma d^2 = 4.5$; $1 - \dfrac{6 \times 4.5}{9 \times 80} = 0.9625$ [M1W1]
(c) Positive correlation; [1]
(d) Mean height $= 216 \div 9 = 24$;
Mean weight $= 38.7 \div 9 = 4.3$ [1]

(e)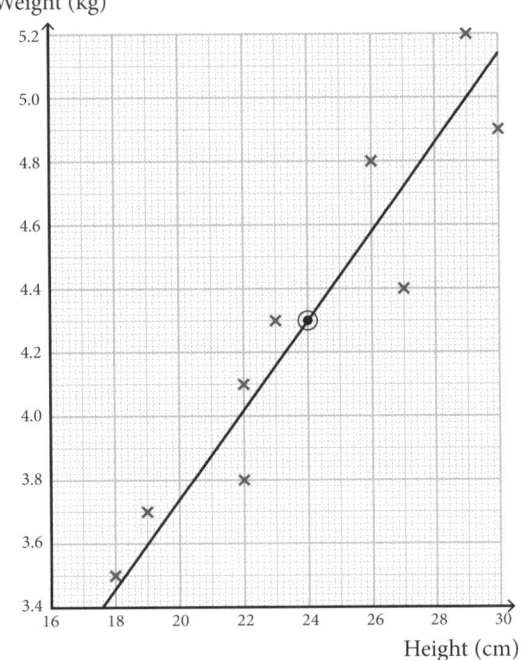

[MW1W1]
(f) Gradient $= \dfrac{5 - 4.3}{29 - 24} = 7/50 = 0.14$; $y = mx + C$;
$4.3 = 0.14 \times 24 + C$; $C = 0.94$;
Equation is $y = 0.14x + 0.94$ [M2W1]

5. (a)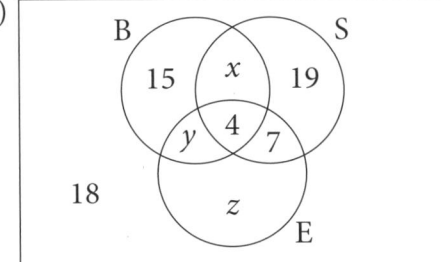

(i) $x = 42 - 19 - 4 - 7 = 12$; P(bacon and sausages)
$= 16/116$; [MW1W2] (ii) $116 - (15 + 12 + 19 + 18) =$
52; P(eggs) $= 52/116$ [1] (b) $11/52$ [M1W1]

6. (a) $z = \dfrac{45.52 - 37.4}{5.8} = 1.4$; P($z > 1.4$)
$= 1 - P(z < 1.4) = 1 - 0.9192 = 0.0808$ [M2W2]
(b) $z = \dfrac{24.93 - 37.4}{5.8} = -2.15$; P($z > -2.15$)
$= 0.9842$ [M1W1]

7. (a) $p^6 + 6p^5q + 15p^4q^2 + 20p^3q^3 + 15p^2q^4 + 6pq^5 + q^6$
[M1W1]
(b) $p = $ P(rains) $= 0.64$; $q = 0.36$ (i) $15p^2q^4$
$= 15(0.64)^2(0.36)^4 = 0.10$ [M2W1] (ii) P(at least 3)
$= 1 - \{P(0) + P(1) + P(2)\} = 1 - \{q^6 + 6pq^5 + 15p^2q^4\}$
$= 1 - 0.128 = 0.87$ [M1W2]

8. P(bungalow and detached house)

Answers

$$= \frac{x}{x+10} \times \frac{10}{x+9} + \frac{10}{x+10} \times \frac{x}{x+9} = \frac{10}{21}$$

so $\frac{20x}{x^2+19x+90} = \frac{10}{21}$; $10x^2 + 190x + 900 = 420x$;

$10x^2 - 230x + 900 = 0$; $x^2 - 23x + 90 = 0$;

$(x-5)(x-18) = 0$; so $x = 5$ (impossible) or 18

P(bungalow and bungalow) = $^{18}/_{28} \times ^{17}/_{27} = ^{17}/_{42}$

[MW3W2]

Revision Exercise 2

1.
Number of points	Number of people (f)	x	fx	fx^2
1 – 5	3	3	9	27
6 – 14	14	10	140	1400
15 – 20	9	17.5	157.5	2756.25
21 – 22	4	21.5	86	1849
Total	30		392.5	6032.25

(a) 13.08 [M1W1] (b) 5.48 [M1W2]

2. (a) All books = $9.60 \times 40 = 384$;
Hardback: $11 \times 28 = 308$;
Paperback = $384 - 308 = 76$;
Mean = $76 \div 12 = £6.33$ [M1W1]
(b) All books: $3.2 = \sqrt{(x^2 \div 40) - 9.60^2)}$
$\Sigma x^2 = 4096$; Hardback: $2.50 = \sqrt{(x^2 \div 28) - 11^2)}$
$\Sigma x^2 = 3563$; Paperback: $\Sigma x^2 = 4096 - 3563 = 533$
$s = \sqrt{(533 \div 12) - 6.33^2)} = 2.09$ [M1W2]

3.
Age	Dog	Cat	Total
<6	13	8	21
≥6	3	4	7
Total	16	12	28

(a) $^8/_{12}$ [M1W2] (b) $^3/_7$ [1]

4. (a) [MW2]

Age (years)	7	5	9	14	10	15	9	9
Cost (£)	13000	14200	8600	7000	9000	6400	11000	10400
Rank (Age)	2	1	4	7	6	8	4	4
Rank (Cost)	7	8	3	2	4	1	6	5
d^2	25	49	1	25	4	49	4	1

(b) $\Sigma d^2 = 158$; $1 - \frac{6 \times 158}{8 \times 63} = -0.88$ [M1W1]

(c) Negative correlation [1]

(d) Mean age = $78 \div 8 = 9.75$;
Mean cost = $79\,600 \div 8 = 9950$ [1]

(e)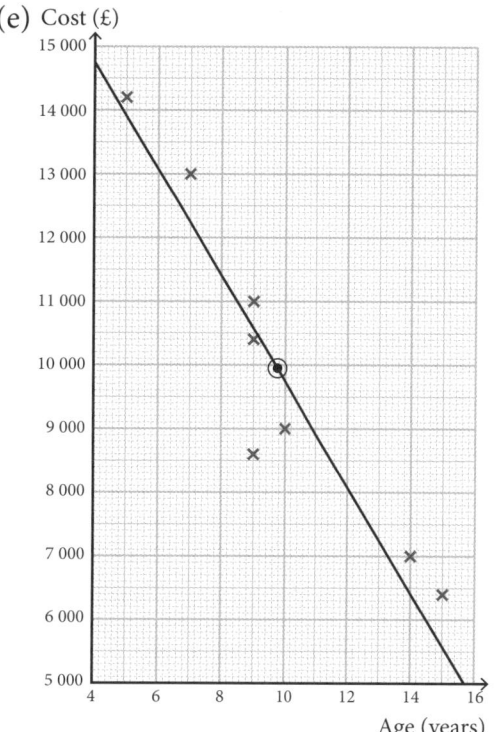

[MW1W1]

(f) Gradient = $\frac{6000 - 9950}{14.5 - 9.75} = -831.6$; $y = mx + C$;

$9950 = -831.6 \times 9.75 + C$; $C = 18058.1$;

Equation is $y = -831.6x + 18058.1$ [M2W1]

5. (a)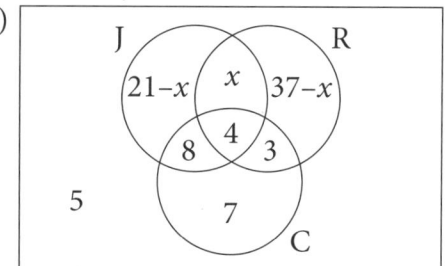

$21 - x + x + 37 - x + 8 + 4 + 3 + 7 + 5 = 76$;
so $x = 9$; $21 - 9 = 12$; P(only Jazz) = $^{12}/_{76}$ [MW1W2]

(b) $^7/_{44}$ [M1W1]

6. (a) $z = \frac{52 - 58}{8} = -0.75$; P($z \leq -0.75$)
$= 1 - P(z \geq -0.75) = 1 - 0.7734 = 0.2266$ [M2W2]

(b) $z = \frac{70 - 58}{8} = 1.5$; P($z \geq 1.5$)
$= 1 - P(z \leq 1.5) = 1 - 0.9332 = 0.0668$ [M1W1]

7. (a) $p^7 + 7p^6q + 21p^5q^2 + 35p^4q^3 + 35p^3q^4 + 21p^2q^5 + 7pq^6 + q^7$ [MW2]

(b) $p = $ P(heads) $= ^3/_5$; $q = ^2/_5$
(i) $q^7 = (^2/_5)^7 = ^{128}/_{78125}$ [M2W1]
(ii) P(4 or 5 or 6 or 7 heads)
$= p^7 + 7p^6q + 21p^5q^2 + 35p^4q^3$
$= ^{11097}/_{15625}$ or 0.710208 [M1W2]

8. P(no laptop sold) = $\frac{12}{x+12} \times \frac{11}{x+11}$

= $\frac{132}{(x+12)(x+11)} = \frac{33}{95}$;

$33(x^2 + 23x + 132) = 132 \times 95$;
$33x^2 + 759x + 4356 = 12\,540$;
$33x^2 + 759x - 8184 = 0$; giving $x = 8$
P(2 laptops sold) = $\frac{8}{20} \times \frac{7}{19} = \frac{14}{95}$ [MW3M1W2]

Revision Exercise 3

1.

Number of texts (x)	Number of people (f)	fx	fx^2
12	7	84	
14	8	112	
17	12	204	
20	x	$20x$	
22	11	242	
26	4	104	
29	2	58	
36	1	36	
Total	$45 + x$	$840 + 20x$	

(a) $\frac{840 + 20x}{45 + x} = 19$; so $840 + 20x = 855 + 19x$; $x = 15$ [M1W1]

(b) 4.81 [M1W1]

2. (a) Mean = $18.7 \times 2 + 3 = 40.4$ minutes [M1W1]
(b) Standard deviation = $2.6 \times 2 = 5.2$ minutes [M1W1]

3. (a) 0.56 = P(a train arriving in Newry is on time and full) ÷ 0.24. So P(a train arriving in Newry is on time and full) = $0.56 \times 0.24 = 0.1344$ [M1W1]
(b) $0.1344 ÷ 0.6 = 0.224$ [M1W1]

4. (a) [MW2]

Volume (cm³)	6	8	5	9	11	12	8	13
Mass (grams)	62	86	64	92	120	138	94	128
Rank (Volume)	2	3.5	1	5	6	7	3.5	8
Rank (Mass)	1	3	2	4	6	8	5	7
d^2	1	0.25	1	1	0	1	2.25	1

(b) $\Sigma d^2 = 7.5$; $1 - \frac{6 \times 7.5}{8 \times 63} = 0.91$ [M1W1]

(c) Positive correlation; [1]

(d) Mean volume = $72 ÷ 8 = 9$;
Mean cost = $784 ÷ 8 = 98$ [1]

(e)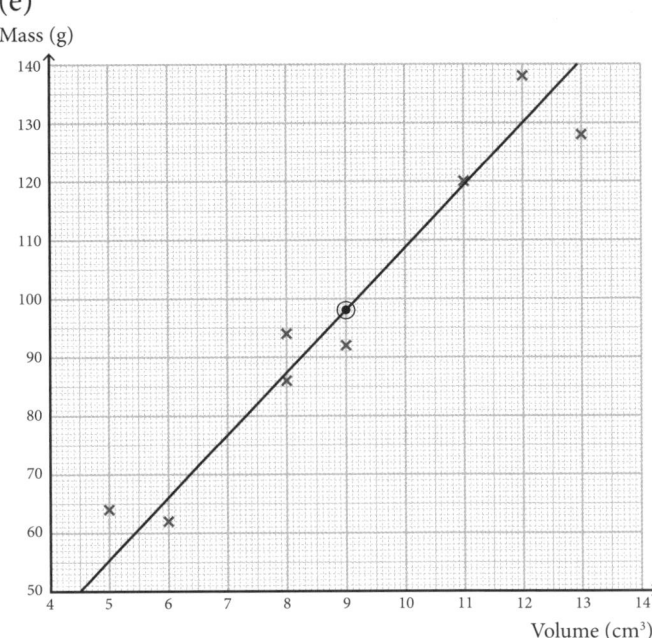

[MW1W1]

(f) gradient = $\frac{130 - 98}{12 - 9} = 32/3$; $y = mx + C$;
$y = 32/3 x + C$; $98 = 32/3 \times 9 + C$; $C = 2$
Equation is $y = 32/3 x + 2$ [M2W1]

5.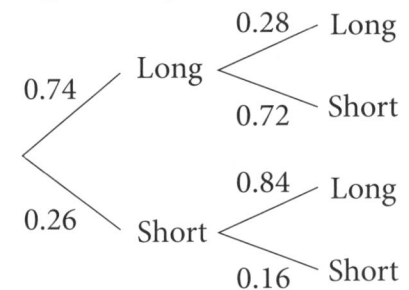

(a) $0.2072 ÷ 0.74 = 0.28$
P(L and S) = $0.74 \times 0.72 = 0.5328$ [M1MW2]
(b) $0.7512 - 0.5328 = 0.2184$
$0.2184 ÷ 0.26 = 0.84$
P(S and S) = $0.26 \times 0.16 = 0.0416$ [M2W1]

6. (a) $z = \frac{44 - 42.912}{2.8} = 0.39$; P($z > 0.39$)
$= 1 - P(z < 0.39) = 1 - 0.6517 = 0.3483$ [M2W2]
(b) $z = \frac{47 - 42.912}{2.8} = 1.46$; P($z \leq 1.46$) = 0.9279 [M1W1]

7. (a) $p^4 + 4p^3q + 6p^2q^2 + 4pq^3 + q^4$ [MW2]
(b) (i) p = P(green) = $10/16 = 5/8$; so $q = 3/8$
$1 - q^4 = 1 - (3/8)^4 = 4015/4096$ [M2W1]
(ii) 1 yellow: p = P(yellow) = $2/16 = 1/8$; so $q = 7/8$
P(1 yellow) = $4 \times 1/8 \times (7/8)^3 = 343/1024$
2 red: p = P(red) = $4/16 = 1/4$; so $q = 3/4$
P(2 red) = $6 \times (1/4)^2 \times (3/4)^2 = 27/128$
P(1 yellow or 2 red) = $343/1024 + 27/128 = 559/1024$ [M1W2]

Answers

8.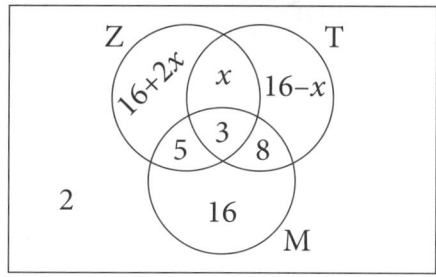

$5 + 3 + x = 8 + x = ⅓$ of those visiting the zoo, so ⅔ of those visiting the zoo = $16 + 2x$;
$16 + 2x + x + 16 − x + 5 + 3 + 8 + 16 + 2 = 74$
giving $x = 4$. 27 went to Titanic; of these $x + 4$ went to the Zoo, so $4 + 3 = 7$ went to the Zoo.
So P(Zoo | Titanic) = 7/27 [MW2M2W3]

Revision Exercise 4

1.

Weight W kg	Number	x	fx
$1 < W \leq 6$	6	3.5	21
$6 < W \leq 14$	15	10	150
$14 < W \leq 17$	22	15.5	341
$17 < W \leq 25$	4	21	84
$25 < W \leq A$	3	x	$3x$

(a) $\frac{596 + 3x}{50} = 13.6$; $x = 28$; so $A = 31$ [M1W1]

(b) 5.88 [M1W1]

2. (a) $p = ½$ [MW1] (b) $54 × ½ + q = 47$, so $q = 20$
3. P(score 60 with 1st or 2nd) = P(score 60 with 1st) + P(score 60 with 2nd) − P(score 60 with both)
= $0.47 + 0.58 − 0.24 = 0.81$
Answer: $1 − 0.81 = 0.19$ [M1W1]

4. (a) [MW2]

Size (litres)	0.8	2.4	1.2	1.6	1.2	2.8	3.2	2.8
Distance (km)	13.2	6.4	14.8	8.8	14.4	8.4	4	7.6
Rank (Size)	1	5	2.5	4	2.5	6.5	8	6.5
Rank (Distance)	6	2	8	5	7	4	1	3
d^2	25	9	30.25	1	20.25	6.25	49	12.25

(b) $\Sigma d^2 = 153$; $1 − \frac{6 × 153}{8 × 63} = −0.82$ [M1W1]

(c) Negative correlation [1]

(d) Mean size = $16 ÷ 8 = 2$;
Mean distance = $77.6 ÷ 8 = 9.7$ [1]

(e)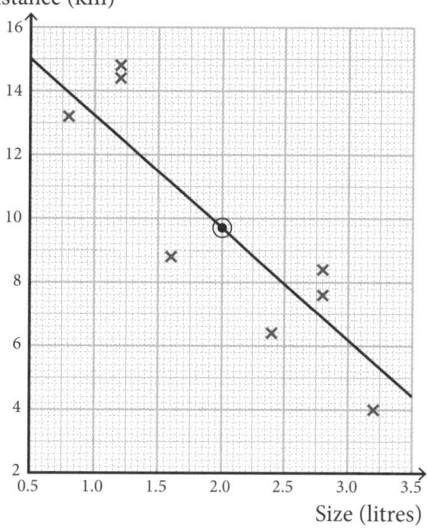

[MW1W1]

(f) Gradient = $\frac{15 − 9.7}{0.5 − 2} = −\frac{53}{15}$ $y = mx + C$;
$y = −53/15 x + C$; $9.7 = −53/15 × 2 + C$; $C = 16 \, 23/30$
Equation is $y = −53/15 x + 16 \, 23/30$ [M2W1]

5. (a)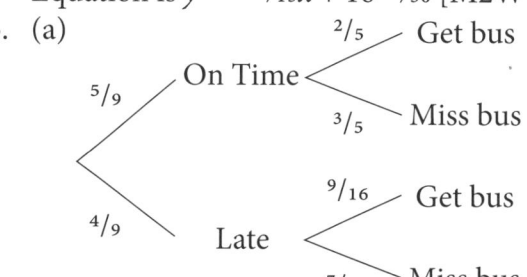

[MW2]

(b) P(on time and get) + P(late and get)
= $5/9 × 2/5 + 4/9 × 9/16 = 17/36$ [M1W1]

(c) ⅖ [1]

(d) P(not on time and get) = $4/9 × 9/16 = ¼$
P(not on time | get) = P(not on time and get) ÷ P(get) = $¼ ÷ 17/36 = 9/17$ [M1W1]

(e) $17/36 × 17/36 × 19/36 + 17/36 × 19/36 × 17/36 + 19/36 × 17/36 × 17/36 = 16473/46656 = 5491/15552$ [M1W1]

6. (a) $z = \frac{2.2 − 2.3}{0.52} = −0.19$; P(z > −0.19) = 0.5753 [M2W2]

(b) $z = \frac{2.76 − 2.3}{0.52} = 0.88$; P(z ≥ 0.88)
= $1 − P(z ≤ 0.88) = 1 − 0.8106 = 0.1894$ [M1W1]

7. (a) $p^8 + 8p^7q + 28p^6q^2 + 56p^5q^3 + 70p^4q^4 + 56p^3q^5 + 28p^2q^6 + 8pq^7 + q^8$ [MW2]

(b) p = P(Lily wins) = 0.52; So $q = 0.48$
(i) $28 × 0.52^2 × 0.48^6 = 0.0926$ [M1W1]
(ii) P(0 or 1 or 2) = $28p^2q^6 + 8pq^7 + q^8 = 0.12$ [M1W1]

8.

Score (x)	Frequency (f)	fx	fx²
1	4	4	4
2	3	6	12
3	2	6	18
n	1	n	n²

Mean = $\frac{16+n}{10}$; Square of standard deviation
= $\frac{34+n^2}{10} - \frac{(16+n)^2}{10^2} = 14$
So: $10(34+n^2) - (16+n)^2 = 1400$;
$340 + 10n^2 - 256 - 32n - n^2 = 1400$;
$9n^2 - 32n - 1316 = 0$;
$(9n + 94)(n - 14) = 0$;
So $n = 14$ [MW2M3W3]

Revision Exercise 5

1.

Length l cm	Number (f)	x	fx
1 < l ≤ 4	2	2.5	5
4 < l ≤ 7	6	5.5	33
7 < l ≤ 10	5	8.5	42.5
10 < l ≤ 13	4	11.5	46
13 < l ≤ 16	x	14.5	14.5x
Total	17 + x		126.5 + 14.5x

(a) $\frac{126.5 + 14.5x}{17 + x} = 8.5$; $x = 3$ [M1W1]

(b) 3.67 [M1W1]

2. (a) (i) 10 people: $12 \times 10 = 120$;
Total = $120 + x + 2x$
Mean = $\frac{120 + x + 2x}{12} = 11.5$; $x = 6$
So David 6 minutes [MW1M1W1]
(ii) Carly 12 minutes [1]
(b) 10 people: $2 = \sqrt{(\Sigma x^2 \div 10) - 12^2}$; $\Sigma x^2 = 1480$
Total $\Sigma x^2 = 1480 + 36 + 144 = 1660$
$s = \sqrt{(1660 \div 12) - 11.5^2} = 2.47$ minutes [M1W1]

3.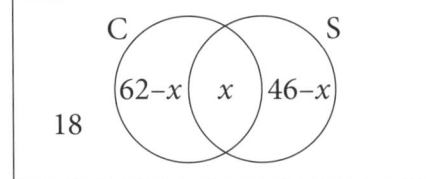

(a) $62 - x + x + 46 - x + 18 = 115$; So $x = 11$
Answer: ¹¹⁄₁₁₅ [M1W1]
(b) ¹¹⁄₆₂ [1]

4. (a) [MW2]

Time (minutes)	49	48	34	45	31	42	43	35	37	39
Speed (mph)	38	52	63	44	71	52	34	52	56	74
Rank (Time)	10	9	2	8	1	6	7	3	4	5
Rank (Speed)	2	5	8	3	9	5	1	5	7	10
d²	64	16	36	25	64	1	36	4	9	25

(b) $\Sigma d^2 = 280$; $1 - \frac{6 \times 280}{10 \times 99} = -0.70$ [M1W1]
(c) Negative correlation [1]
(d) Mean time = ⁴⁰³⁄₁₀ = 40.3
Mean speed = ⁵³⁶⁄₁₀ = 53.6 [1]
(e)

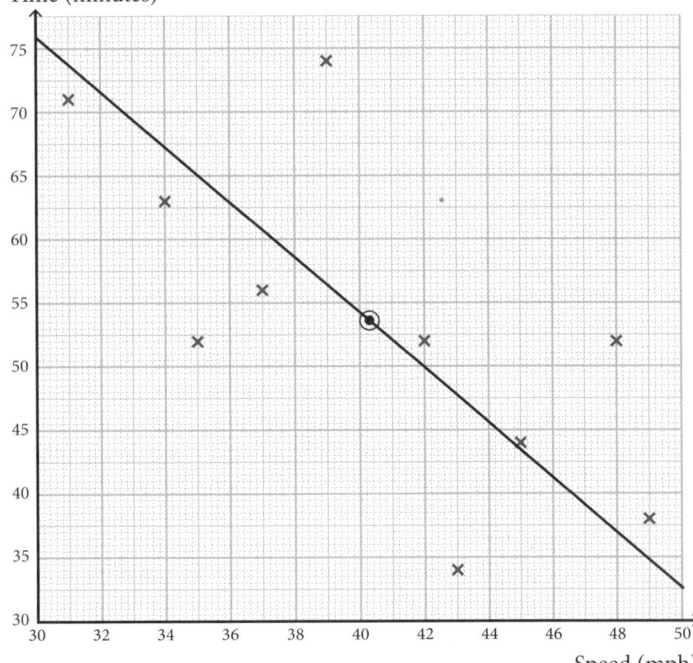

[MW1W1]
(f) gradient = $\frac{76 - 53.6}{30 - 40} = -2.24$; $y = mx + C$
$y = -2.24x + C$; $53.6 = -2.24 \times 40.3 + C$;
$C = 143.872$;
Equation is $y = -2.24x + 143.872$ [M2W1]

5.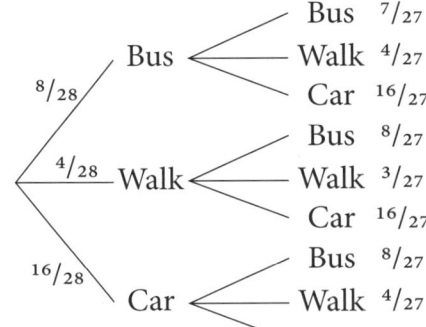

(a) (i) P(bus and bus or walk and walk or car and car) = ⁸⁄₂₈ × ⁷⁄₂₇ + ⁴⁄₂₈ × ³⁄₂₇ + ¹⁶⁄₂₈ × ¹⁵⁄₂₇ = ¹¹⁄₂₇
[MW1M1W1]
(ii) P(at least 1 walk) = 1 − P(1st doesn't walk and 2nd doesn't walk) = $1 - \frac{24}{28} \times \frac{23}{27} = \frac{17}{63}$ [M1W1]
(b) ¹⁶⁄₂₇ [1]

6. (a) $z = \frac{240 - 210}{24} = 1.25$; P(z < 1.25) = 0.8944
[M2W2]

Answers

(b) $z = \dfrac{145 - 210}{24} = -2.71$; $P(z < -2.71)$
$= 1 - P(z > -2.71) = 1 - 0.9966 = 0.0034$ [M1W1]

7. (a) $p^5 + 5p^4q + 10p^3q^2 + 10p^2q^3 + 5pq^4 + q^5$ [MW2]
   (b) (i) $p = P(\text{watches TV}) = 0.61$; $q = 0.39$
   So Answer $= 0.61^5 = 0.084$
   (ii) $1 - 0.61 = 0.39$; $0.39 \div 3 = 0.13$; $0.13 \times 2 = 0.26$;
   So: $p = P(\text{play outside}) = 0.26$; $q = 0.74$;
   $10p^2q^3 = 0.27$ [MW1M1W1]
   (c) $P(\text{homework}) = 0.13$; $q = 0.87$
   $P(\text{at least } 1) = 1 - q^5 = 0.50$ [M1W1]

8. Let n be the number of boys who swam 5 lengths.
   Boys:

Number of lengths (x)	Number of boys (f)	fx
5	n	$5n$
10	17	170
15	14	210
20	9	180
25	8	200
Total	$48 + n$	$760 + 5n$

   Mean $= \dfrac{760 + 5n}{48 + n}$

   Girls:

Number of lengths (x)	Number of girls (f)	fx
5	$6n$	$30n$
10	18	180
15	15	225
20	12	240
25	18	450
Total	$63 + 6n$	$1095 + 30n$

   Mean $= \dfrac{1095 + 30n}{63 + 6n}$

   $\dfrac{760 + 5n}{48 + n} = \dfrac{1095 + 30n}{63 + 6n}$;

   $30n^2 + 4875n + 47880 = 30n^2 + 2535n + 52560$;
   $2340n = 4680$;
   $n = 2$
   So number of girls who swam 5 lengths $= 6n = 12$
   [MW5]

Meeting the requirements of the CCEA GCSE Further Mathematics specification, this is one of four revision booklets that address the course. This workbook is set out in the form of five revision tests and covers all of the elements of the *Statistics* module.

These valuable questions were specially commissioned for the booklet and are not from past papers. Full answers are included at the rear and contain not only the final answer but, where appropriate, an indication of the process required to reach the given solution. The workbook has been through a meticulous quality assurance process by a GCSE Mathematics expert.

What workbooks do I need?

Students sitting CCEA GCSE Further Mathematics will usually need four workbooks – *Pure Maths 1* and *Pure Maths 2*, plus the *Mechanics* workbook and the *Statistics* workbook.

£3.99

www.colourpoint.co.uk